A Soldier's

POCKET THERAPIST

By

Ginny Sartini, MSW, LCSW

Cover design by Mary Kramer http://www.Lulu.com/TreeTopLeaf.

This book is designed as a source of reference to help prevent mental health deterioration while on active duty. If you have a psychiatric diagnosis, expert assistance is required and the services of a competent professional should be sought.

CONTENTS

Page

Preface.. 6

How To Use This Booklet............................ 7

Homesickness... 9

Anxiety/Worry.. 11

Depression/Sadness/Irritability....................... 13

Anger.. 15

Stress Management.................................. 16

Sleep Disturbance/Dreams and Nightmares......... 18

Grieving Losses....................................... 20

Witnessing/Experiencing a Traumatic Event....... 23

Handling Flashbacks of Trauma.................. 25

Dealing With Having Killed Another............... 27

Thoughts of Suicide................................ 29

Military Sexual Trauma............................ 30

Alcohol and Substance Abuse...................... 31

Appendix A: Questions for Journaling............ 34

Appendix B: Progressive Relaxation Technique... 36

Appendix C: Deep Breathing...................... 38

Appendix D: Guided Imagery..................... 40

Appendix E: Meditation........................... 42

Appendix F: Working with Negative Feelings.... 44

Appendix G: Grounding Techniques............. 45

Appendix H: Need for Medication............... 47

Appendix I: Online resources...................... 49

References... 50

About the Author 51

**Many thanks to my sister, Bobbie Sartini, for editing the book
and to Mary Kramer
for designing the cover.**

Dedicated to our Heros

who relive the traumas of war every day in their minds.

May they come home soon.

PREFACE

I have read with great sadness several articles that discussed the mental health of soldiers in Iraq and Afghanistan. They discussed low morale, high stress, and gaps in the soldier's support system. From April to December 2003 there were 23 soldiers who killed themselves in Iraq and Kuwait. At least seven other soldiers committed suicide last year after returning to the United States.

On March 26, 2004 the Associated Press released the results of an Army survey that reported shortages of anti-depressant medications and sleeping drugs, inadequate training in combat stress control, and ill-defined standards of care in Iraq and Kuwait. Three-fourths of those who had participated in this survey said that they had received no help at any time in Iraq from a mental health professional, a doctor, or a chaplain.

A Soldier's Pocket Therapist was developed as an attempt to help with shortages of support personnel, to possibly prevent the loss of young lives to suicide, to provide help for those who suffer in silence for fear of hurting their career, and to help bring soldiers home mentally healthy.

A Soldier's Pocket Therapist was my way of giving something back to my country and to support those who give so generously for our freedoms.

How To Use This Booklet

You cannot control what happens to you, but you can control how you respond to it. Aside from our genetic makeup, it is not what we go through in life, but the support system we have in place at the time of stress and our ability to process our feelings, that affect our mental health. To this end, *A Soldier's Pocket Therapist* was developed as a tool to aide soldiers in processing their feelings therapeutically. It is not meant to be read from cover to cover. Rather it is meant to be a book of reference in times of need.

Step 1: Choose the issues you are dealing with from the Table of Contents.

Step 2: Locate the corresponding page number for those issues.

Step 3: Review the relevant interventions on the assigned page and select those that you would like to try and for which you have the tools to complete the task. You may choose to do one or all of the tasks. However, it is advisable to work on those that evoke a lot of feelings one at a time with a day or so between them so that you are able to process the feelings that arise from the task. Stuffed feelings can cause problems later in life.

Step 4: Review any appendix referenced and complete the exercises assigned to them also. If an exercise calls for writing something out, it is very important that you do so in order to process your feelings.

Step 5: Journal. Journal. Journal.

A FINAL NOTE:

An important aspect in processing your feelings is that each task should be tailored to your individual problems and needs. Some tasks listed may not be beneficial to you. If that is the case, it is suggested that you try another task. However, don't permanently discard any of the tasks because you feel uncomfortable doing them. Remember, "No pain – no gain." If none of the tasks are effective as stated, perhaps the list will give you other ideas that would be beneficial for you to try. *The point is to maintain good mental health in any way that works for you.*

IF PROBLEMS PERSIST, PLEASE CONTACT A PROFESSIONAL.

HOMESICKNESS

1. Relaxation Techniques (Appendices B & C).

2. Guided Imagery (Appendix D).

3. Participate more frequently in peer-group activities.

4. Get a hobby that reinforces self confidence (write a book, make toys to hand out to kids, adopt an elementary school class in the United States and e-mail them about the culture, etc.).

5. List your fears associated with your separation from family and friends. Differentiate between realistic and unrealistic fears.

6. Process feelings of abandonment, separation, grief, sadness, etc. (Appendix F).

7. Explore your fears associated with you and your family being apart and where that fear comes from.

8. List your life experiences of emotional and physical abandonment.

    a. Emotional abandonment is when others are physically present, but don't talk, share themselves, or listen to the family member needing emotional nurturing.

    b. Physical abandonment is when basic physical needs are not met or are met by the children themselves.

9. List unmet needs since leaving home and ways to meet those needs yourself.

10. Change the negative ways you look at the feeling of homesickness to a positive way to look at it. For example, homesickness gives you purpose, and purpose keeps you going and helps you to survive.

11. Find YOUR purpose for being where you are and "bloom where you are planted."

12. Write letters home, draw pictures to share with family, maybe even write a storybook to send home to your children to help them understand your absence.

ANXIETY/WORRY

You'd have to be dead not to feel anxiety. We all do. Remember: "In life pain is inevitable; suffering is optional" (Peurifoy, 2005).

1. Guided Imagery (Appendix D).
2. Relaxation Techniques (Appendix B).
3. Physical exercise.
4. Breathing (slow it down) (Appendix C).
5. Self talk:
   a. Replace "What if" with "So what if."
   b. Use "I like, I want, I have decided" instead of "can't, always, never, should or must."
   c. Don't criticize people (including yourself); rather, criticize the behaviors.
   d. Ask, "What evidence do I have to support my assumption(s)? Is it/are they realistic?"
6. Process the feeling of fear (Appendix F).
7. Focus away from anxiety symptoms. Distract yourself
   a. Count backwards from 1,000.
   b. Work a crossword puzzle.
   c. Journal (Appendix A).
8. List key unresolved life conflicts and work one-by-one towards resolution.
9. Avoid caffeine in coffee, sodas, medications, and foods such as chocolate.
10. Exercise three times a week for at least 20 minutes.

11. Thought Blocking – Whenever you find yourself obsessing over an unwanted thought or situation either scream "Stop" out loud (preferably) or inside your head and immediately switch your thoughts to something more pleasant. You may have to do this many times, but it can work if you stick to it.

12. Breathe into a paper bag and slow your breathing down.

13. See your limitations as being human and not as being inadequate.

14. Listen to relaxing music.

15. Remember, it is not people and events that generate your emotions, but the meaning you give to people and events.

DEPRESSION, SADNESS AND/OR IRRITABILITY

1. Eat nutritiously – avoid sugar, caffeine, and fatty foods. Eat foods high in complex carbohydrates (whole grain breads, pasta, rice, potatoes, and vegetables).
2. Get needed sleep.
3. Get a massage.
4. Make a list of positives in your life and read the list often.
5. Get physical exercise daily (play sports, take a walk).
6. Increase social involvement.
7. Feelings get exaggerated and/or magnified with depression. Express feelings through journaling (Appendix A).
8. Focus outside yourself. Find purpose to your life now and in the future.
9. Draw, paint, work with clay, write poetry, etc. to express your feelings.
10. Spend time with children.
11. Come up with new ideas to add to this book to help others.
12. Make a list of your accomplishments and refer to it when feeling discouraged.
13. Talk with someone.
14. Spend time with positive, non-judgmental friends.
15. Create/attend and participate in a support group of your buddies.
16. Spend time with a pet.
17. Remember that depression ends eventually.
18. Live one day at a time and complete one thing at a time.
19. Find humor in all that you do.
20. Get out in the light daily.
21. Keep busy.

22. Balance rest and activity, work and play, responsibility and relaxation.

23. Pray or meditate (Appendix E).

24. Make a list of things you enjoy doing and refer to it when you are feeling down.

25. Depression can be caused by a chemical imbalance that is triggered by stress. Medication can help you to function effectively again (Appendix H).

26. Force yourself to change negative thoughts to positives thoughts.

27. Listen to uplifting music; read a funny book.

28. Use daily affirmations.

29. Use Relaxation Techniques (Appendix B) and Guided Imagery (Appendix D).

30. Vitamins B-6 and B-12 can help with depression.

31. Remember, some people are put in our lives for a short time to meet certain needs. If the relationship doesn't work out it is because you are being prepared to meet the next person who can meet your future needs.

ANGER

Anger is a secondary emotion and is used to protect us. It is important to work through the anger so that you can identify the primary feeling(s) triggering your anger. You can do this by:

1. Identify who or what you are really angry at.
2. Do feeling work (Appendix F) with anger.
3. Beat a pillow, hit a punching bag or other item that is not harmful.
4. Shred newspapers by hand.
5. Scream with your mouth open or shut or into a pillow.
6. Play sports.
7. Work with clay/mud.
8. Jog.
9. Make a list of experiences in your life that have hurt and led to anger.
10. Write therapeutic letters (not for mailing) directed to those who have harmed you focusing on the reasons for your anger.
11. Make a list of socially acceptable ways to handle anger and find one that works for you.
12. Identify what feeling is fueling your anger and do feeling work (Appendix F) with that feeling.
13. List the ways your anger has had a negative impact on your life.
14. Write a letter to the non-tangible source of your anger, i.e., the war, the enemy as a whole, death, God, etc.
15. Identify your unmet needs and then list how you can meet those needs yourself.

STRESS MANAGEMENT

1. Exercise.

2. Rest.

3. Take a nap.

4. Meditate (Appendix E).

5. Do relaxation techniques (Appendix B & C).

6. Take a 15-minute break.

7. Listen to music.

8. Keep decision making to a minimum.

9. Make a list of things that bring you comfort and read the list when you are stressed.

10. Support each other through teamwork.

11. Diaphragmatic breathing – suck in your abdomen while inhaling and relax while exhaling.

12. Normal breath rate is 8 to 16 breaths per minute at a relaxed state. Time yours and adjust.

13. Plan ahead as much as possible so you aren't rushed at the last minute. Manage your time as much as possible.

14. Eat nutritiously; drink plenty of water; eat fiber; use low-fat dairy products; maintain a healthy weight; decrease fat, sugar, and salt intake; decrease caffeine and alcohol consumption, substance abuse and smoking.

15. Eat food high in Vitamin B, especially B6 and B12 (whole grains, nuts, seeds, and beans).

16. Develop a support system – Talk out your problems with a good listener.

17. Do something you really enjoy.

18. Take action to settle matters that are troubling you.

19. Work through any anger you feel (See Chapter on Anger).

20. Be willing to compromise.

21. Handle one thing and one day at a time.

22. Keep a positive attitude – Don't dwell on the negatives.

23. Minimize the "unknowns" as much as possible.

24. Have a reason and goal for living.

25. Pay attention to your feelings. Do feeling work if necessary (Appendix F).

26. List the things that are causing you stress. Cross off items as they become resolved. This way you can look back, see improvement, and know that everything works out sooner or later.

27. Visualize being in control and being successful.

28. Ask for help.

29. Find a way to turn negatives into positives.

30. Maintain a sense of humor; spend time with others who do the same.

31. Spend time with a child or a pet.

SLEEP DISBURBANCE, DREAMS AND NIGHTMARES

1. Relate your dreams to life stressors and feelings. For example: You dream you are in a very large shopping mall and can't find the exit. Ask: "What do I feel in the dream"? Trapped! What do you feel in your life at the present time? Trapped! Know that your unexpressed feelings are processed at night through your dreams. After identifying the feeling(s) you may want to do feeling work (Appendix F).

2. Journal before going to sleep (Appendix A). Include daily stressors.

3. Don't nap during the day.

4. Use Relaxation Techniques (See Appendices B, C, D & E).

5. Count **backwards** from 1,000. If your mind starts to drift to daily stressors in your life say, "Stop" and think of something pleasant. Do this over and over until you can sleep.

6. Don't use alcohol or substances to sleep unless prescribed by your doctor.

7. Exercise daily, but not before bedtime.

8. Reduce stimulating activity prior to sleep.

9. Don't eat heavy foods within two hours of going to bed.

10. Medications may be needed. (Appendix H).

11. Don't use your bed for any other purpose but to sleep.

12. If you wake up and can't get back to sleep or can't get to sleep, don't lie there and toss and turn. Read a book, journal about what is bothering you or write letters to those upsetting you (not necessarily for mailing); or do some other activity that makes you sleepy again. Even if you don't feel sleepy after any of these exercises, lay down and see if you can fall asleep.

13. Listen to soft, slow music such as light jazz or orchestral pieces of 60 to 80 beats a minute.

14. If you are experiencing bad dreams or nightmares, before you go to sleep every night, tell yourself that if you dream a dream of trauma you will find a way to be in control of the event and be victorious.

15. Keep a journal of your dreams and look for a common theme/feeling among your dreams, then relate it to your current life situations.   Once any feelings are identified, do feeling work (Appendix F).

GRIEVING LOSSES

People grieve more than the death of a loved one. We can grieve the loss of a job or the loss of life as we know it. We can even grieve the loss of an abused substance when we stop an addiction. Some helpful suggestions for grieving include the following:

1. Face the crisis actively so as to realize the full reality of what has happened and to accept the permanency of the loss. Although it is painful, it is this pain that activates the healing process.

2. Normalize your feelings by knowing the stages of grief, which are:

      a. Shock and Denial

      b. Bargaining

      c. Anger

      d. Blame/guilt

      e. Depression/Remorse

      f. Acceptance

3. Allow yourself time to grieve. Feel the feelings. It is okay to cry. It is a normal reaction.

4. Do feeling work (Appendix F) about your feeling of loss, abandonment, sadness, etc.

5. Visualize the person lost and say what you would like to have said but couldn't or didn't (regrets and disappointments as well as the positives of your relationship with him/her).

6. Identify what the loss will mean to you in the future and list positive ways in which you can cope with the loss.

7. Let go – write a therapeutic letter in your journal (not for mailing) to bring closure to the relationship and describing your feelings. Don't forget to address the "if only's," regrets, and disappointments.

8. Wait a day or so after writing your letter, and then write the letter you think you would get back from the person you lost.

9. Draw pictures of the story of the loss.

10. Draw pictures of your feelings (Appendix F).

11. Make a list of ways this loss has impacted your life.

12. Journal daily about the loss to get through the stages of grief. Don't be afraid to list the things you do not miss about the person, as anger is a stage of grief that can be worked through in this manner.

13. Write a list of positive and negative things about the person lost – reflection.

14. Release repressed anger constructively (See chapter on anger).

15. Talk to someone who understands.

16. Keep your own sense of purpose in life.

17. Don't make hasty decisions in response to your loss.

18. Ask for specific help; don't wait for others to figure out what you need.

19. Remember that grieving a death continues long after the funeral. However, an end to grief does not mean you stopped caring about the person. Love lasts beyond grief through a commitment to living your own life fully.

20. Consult with professionals if grief becomes intense and prolonged.

21. Pray the serenity prayer: "God, grant me the serenity to accept the things I cannot change, courage to change the things I can, and the wisdom to know the difference. Oh divine master, grant that I may not so much seek to be consoled as to console, to be understood as to

understand and to be loved as to love. For it is in giving that we receive; it is in pardoning that we are pardoned; and it is in dieing that we are born to eternal life."

22. Hold a mock funeral, wake, or memorial service.

23. Light a candle or find another symbol for honoring their memory.

24. List your fears regarding your own death.

25. Write poetry.

26. Make a memorial book of those lost and send it to their survivors.

WITNESSING/EXPERIENCING TRAUMA

1. Write down the details of the trauma as you experienced them as soon as possible. Recall the sights, the sounds, the smells, and touch. Record your role in the event and your emotional reactions. Read it over and over again to desensitize yourself to the trauma, then put it away in a safe place to save it for when and if these memories again become a problem.

2. If possible, make an audio recording of your descriptions of the event. Play it over and over again to desensitize yourself to the trauma, then put it away in a safe place to save it for when and if these memories again become a problem.

3. Talk about the trauma with others sharing your experience of the events.

4. If you don't feel ready to process these memories, use Meditation (Appendix E) or Guided Imagery (Appendix D) to picture yourself placing the traumatic event in a container and hiding it away until you are ready to deal with it.

5. Maintain a social support system.

6. Direct your anger where it belongs: on the event, not on the person (see chapter on anger).

7. Make a list of the impact the events have had on your life.

8. Process your feelings (Appendix F).

9. Validate your fears but focus on the facts.

10. Normalize your feelings by knowing the stages of response to trauma (Horowitz):

> Outcry phase – acute alarm to stunned inability to take in the meaning of the experience.

Denial phase – includes amnesia, numbness, sleep disturbances, physical symptoms, frantic over-activity and withdrawal.

Intrusive phase – exaggerated startle responses, intrusive thoughts, preoccupation, indifference, chronic arousal state, sleep and dream disturbances, fear of going insane (this one is very important to remember as a normal phase and not necessarily a true outcome).

Working through phase – examines meanings of the traumatic event, mourns losses and injuries, and considers new plans for coping with failure.

Completion phase – recognizes the impact of the trauma on your psyche, exhibits hopeful plans for the future and resumes work and leisure activities.

NOTE: Abnormal reactions to trauma include withdrawal, substance abuse, dissociative states, psychoses, and depressive reactions. If you experience these symptoms and they do not resolve within a few weeks, seek professional help.

11. Journal (Appendix A).

12. Explore past traumas in your life. Trace your feelings back in time to other events that have caused you to feel the same way and process all issues connected to this feeling through letter writing, journaling, etc.

13. If you are experiencing survivor's guilt, look at it from a different perspective. For example, if you had lost your life instead, your family and friends may be less able to handle it than the family of the deceased. Maybe you have not yet fulfilled your purpose in life.

14. Flashbacks and intense recollections of the event(s) can be treated with grounding techniques (Appendix G).

15. Use Meditation (Appendix E).

HANDLING FLASH BACKS OF TRAUMA

1. Find grounding techniques that work for you (Appendix G).

2. Relaxation Techniques (Appendices B & C).

3. Thought blocking: place a rubber band on your wrist and snap it or yell out "stop" or scream it in your head when you feel yourself going deep into the memories.

4. Replace negative thoughts with positive thoughts. Seek pleasure, not pain.

5. Eat right, get ample rest, exercise, and be involved in leisure activities.

6. List sources of emotional pain, feelings of fear, inadequacy, rejection or abuse and avoid them if possible.

7. Prepare for triggers by noticing your body's reactions prior to flashbacks and remember them so that you can stop future attacks before they occur.

8. List the things you tell yourself about the event that create a negative response. Share these thoughts with someone you trust.

9. Recreate the trauma in your mind using guided imagery (Appendix D) but reframe the memory by adding a different ending with which you feel comfortable.

10. Keep track of the symptoms that develop following exposure to the trauma.

11. Get important people in your life to help you with your recovery.

12. Tell others and get involved with others who are supportive.

13. Keep a journal of the recurring images or memories that are associated with the trauma.

14. Meditate (Appendix E).

15. It has been proven that witnessing a traumatic event can change the chemistry in our brains. Medications have been proven successful in treating this chemical imbalance (Appendix H).

16. Anger is a valid feeling. Distinguish between constructive and destructive anger. Work to turn destructive anger to constructive anger.

17. Learn to love yourself. Show yourself compassion and respect.

DEALING WITH HAVING KILLED ANOTHER
(Kilner)

1. Do feeling work with the feeling of guilt (Appendix F). At the end of the exercise, guilt can be destroyed or made your friend. In making it your friend, tell yourself that the ability to feel guilt and remorse is the main thing that separates you from an antisocial personality. In that respect, guilt is a healthy response, not a weakness, to having killed.

2. Know that it is okay to feel guilty. It doesn't mean you are.

3. In a sense, we have all killed. Whenever we gossip about another we have killed their reputation.

4. Know that God judges us by the intentions of our heart. What were your real intentions?

5. Differentiate between murder and killing.

6. Know that God made the rules but knew we couldn't keep them. That is why he sent a Savior to meet His own demands.

7. When we experience road rage, we view the opposition as an object (a car). It is not until we see the person behind the wheel that the human side is seen. In war it may help to see the battle as weapon against weapon.

8. Know that the "professionals" have already tried peaceful ways to resolve the conflict and war was the last resort.

9. Know that feelings of guilt about killing in combat are healthy, normal and very common.

10. Talk about your experiences with those who understand. Receive their acceptance, forgiveness, and support.

11. Pray for the enemy.

12. Write a letter to all those you have had to kill. Wait a couple of days and write a letter you feel you would have gotten back as a response to your letter. Keep in mind that the other person probably felt the same about war as you do.

13. Do grief work (see chapter on Grieving Losses).

14. Consult a Chaplain.

THOUGHTS OF SUICIDE

1. Think about what losses would be incurred by others. Don't just assume that you know their responses. Ask them. You may be surprised at how they really feel.

2. Remember that life has its ups and downs. If you stop now, you will never know what could have been. Sooner or later things always work out and mostly for the better.

3. Think about desires for the future.

4. Think about what you still value in life.

5. Remember if you die, the terrorists win.

6. Learn how others problem solve and choose an appropriate coping strategy that fits you.

7. Eat nutritiously – avoid sugar.

8. Get needed sleep.

9. Make a list of positives in your life.

10. Get physical exercise daily.

11. Increase social involvement.

12. Express feelings through Journaling (Appendix A).

13. Focus outside yourself. Find purpose to your life now and in the future.

14. Do anger work (see chapter on anger).

15. Talk to someone honestly. They may have an answer.

16. Stress can cause depression and depression can cause you to loose the ability to put things in proper perspective. Medication can help you to feel normal again (Appendix H).

MILITARY SEXUAL TRAUMA (MST)

1. Remember, you did not invite the MST and you did nothing to deserve it.

2. Confide in someone about your feelings about the MST.

3. List how the MST has affected your life and health.

4. Seek counseling when ready – not because you are at fault, but to learn how to cope with the trauma.

5. If you are feeling suicidal, read the chapter on suicide.

6. Don't cope by abusing alcohol and/or other substances; it will only cause you more trauma (see chapter on Alcohol and Substance Abuse).

7. If you are experiencing flashbacks or intense recollections of the abuse that interfere with your functioning, see the chapter on Handling Flashbacks of Trauma.

8. Know that survivors of MST are more susceptible to depression, substance abuse/use, panic and anxiety disorders, and obsessive-compulsive disorders. Seek medication (Appendix H) if these become a problem.

9. Relaxation therapy (Appendix B).

10. Deep Breathing (Appendix C).

11. Write down all that you remember of the event.

12. Process your anger and other feelings regarding the event (see chapter on Anger and/or Appendix F).

13. If you feel safe enough, report the MST.

ALCOHOL AND SUBSTANCE ABUSE

1. If you have problems you can't cope with before you get high, you will have those same problems plus problems from the substance if you choose to get high. Using substances to cope will only delay the inevitable task of facing the real issues.

2. Many substances are depressants and will only make matters worse in the long run.

3. List the negative ways substances have negatively impacted your life either directly or indirectly.

4. List the negative ways your current living situation encourages you to use. Find realistic alternatives for as many of these as possible.

5. Think about what life would be like without the substance and list the positive effects.

6. Keep a list of things and feelings that trigger your cravings for substances and list alternative ways of avoiding/coping with them.

7. List changes that you can realistically make to keep you from using again.

8. Find or create a support system that does not use substances.

9. List family issues that contribute to your desire to use substances. Determine how these issues make you feel and address those feelings using techniques in Appendix F.

10. Make a list of healthy activities you can do in your free time, like working in this book.

11. Write letters or ask others around you how your substance abuse has negatively impacted their lives.

12. Giving up any substance may require a grieving process. Use the chapter on Grieving Losses to process those feelings.

13. Identify and journal your feelings, behaviors, and attitudes that trigger your desire to use.

14. Look back through your life and identify the feelings you had prior to each substance abuse. Trace that feeling back through your life and identify alternatives that you could have used to cope.

15. List people, places, music, etc. that trigger you and need to be avoided.

16. Learn to relax. (Meditate (Appendix E), Relaxation Techniques (Appendix B), or Guided Imagery (Appendix D)).

17. Make a list of things that bring you comfort and read the list when you are stressed.

18. Develop a support system – Talk out your problems with a good listener.

19. Do something you really enjoy.

20. Handle one thing and one day at a time.

21. Find healthy ways to reward yourself. Ask others for suggestions.

22. Find healthy ways to repent of past wrongdoing.

APPENDICES

APPENDIX A
QUESTIONS FOR JOURNALING
(Author Unknown)

Not all questions will apply to your current situation. Complete the ones that do.

1. What made me upset today?
2. What made me happy today?
3. What did I learn today?
4. What am I grateful for today?
5. What is really going on with me today?
6. What am I really feeling right now?
7. What have I done today that I did not want to do?
8. What caused me to do it?
9. What have I not done today that I really wanted to do?
10. What prevented me from doing it?
11. How is this problem (depression, anxiety, being suicidal, etc.) serving me?
12. What does it allow me to do or not do?
13. What is my motive for staying stuck?
14. What am I avoiding?
15. What do I want?
16. What do I feel deprived of?
17. What am I really afraid of?
18. What changes do I need to make to feel better?
19. What prevents me from making those changes?
20. Am I just unwilling to make the necessary changes?
21. Do I think something or someone ought to change before I do?

22. With whom or what am I really angry?

23. What is preventing me from dealing with my anger?

24. What would it take to release that anger?

25. What are my grudges and grievances against people who have hurt me?

26. Do I feel that I was unfairly treated?

27. Do I want someone to pay (revenge) for hurting me?

28. What secrets am I hiding and from whom am I hiding them?

29. What do I want from life?

30. What am I doing to get it?

31. What or who is standing in my way?

32. Do I want someone to come to my rescue?

APPENDIX B
PROGRESSIVE RELAXATION TECHNIQUE

<u>Basic Technique:</u> Separately tense the individual muscle groups listed below. Hold the tension about five seconds. Release the tension slowly and, at the same time, silently say, "Relax and let go."

A. Head

        Wrinkle your forehead.

        Squint your eyes tightly.

        Open your mouth widely.

        Push your tongue against the roof of your mouth.

        Clench your jaw tightly.

B. Neck

        Push your head back into the pillow or chair.

        Bring your head forward to touch your chest.

        Roll your head to your right shoulder.

        Roll your head to your left shoulder.

C. Shoulders

        Shrug your shoulders up as if to touch your ears.

        Shrug your right shoulder up.

        Shrug your left shoulder up.

D. Arms and hands.

        Hold your arms out and make a fist with each hand.

        One side at a time: Push your hands down into the chair, bed, floor.

        One side at a time: Make a fist, bend your arm at the elbow, and tighten up your arm while holding the fist.

E. Chest and Lungs

> Take a deep breath.
>
> Tighten your chest muscles.
>
> Arch your back.

F. Stomach

> Tighten your stomach area.
>
> Push your stomach area out.
>
> Pull your stomach area in.

G. Hips, legs, and feet

> Tighten your hips.
>
> Push the heels of your feet into the floor, bed, etc.
>
> Tighten your thighs by extending your legs.
>
> Curl your toes under as if to touch the bottoms of your feet.
>
> Bring your toes up as if to touch your knees.

Note: For a quick way to relax, tense all your muscles at the same time, hold them as long as you can, then let go.

APPENDIX C
DEEP BREATHING
(Florida Hypnosis and Counseling Center, Inc.)

Place yourself in a comfortable sitting position with hands on thighs.

Breath 1:

- Take a deep, deep breath inhaling through your nose. Hold it for about 10 seconds. Exhale slowly through a somewhat restricted mouth opening. Allow the body to deflate as you do. Really go limp!
- Resume normal breathing. Focus your sensory attention on the two tension muscles in your jaw and let them relax.
- When you can feel the tension draining out of those muscles and when they are relaxed, take your next deep breath.

Breath 2:

- Take another deep breath as in Breath 1. This time allow body to deflate deep into your abdomen. Resume normal breathing and focus your attention on those two strong muscles that run up either side of the back of your neck.
- Let the tension drain out, and when the muscles are relaxed go on to Breath 3.

Breath 3:

- Take another deep breath. Exhale slowly, and as you do, imagine that you have just taken off a heavy backpack. Let your shoulders feel the relief and relax and sag.
- When your shoulders are relaxed, go to Breath 4.

Breath 4:

- Take your fourth deep breath, and as you exhale, let your arms deflate all the way down to your hands and fingers. Permit your hands and fingers to become very relaxed and as limp as possible.
- When your hands and fingers are limp...

Breath 5:

- Take your fifth deep breath. As you exhale, listen for the internal hum or buzz that only you can hear and let that sound take you to your special, safe, and secure place (Appendix D - optional).

APPENDIX D
GUIDED IMAGERY

Examples will be in [   ]. You may use the example or create your own.

1. If you are tense you may want to close your eyes and do progressive relaxation (Appendix B) or deep breathing (Appendix C) first.

2. Close your eyes and picture a path that leads from your mind into a [peaceful, beautiful, valley] safe place that no one can enter without your permission.

3. Build in and view the safety factors that are in your safe place [the valley is surrounded by high mountains and there are angels standing guard everywhere to prevent anyone from coming in].

4. Picture a spot in your safe place where you can sit/lie and relax [to your right is a bunch of shade trees with a hammock strung between two trees. There is a babbling brook that runs beside it]. Go there and lie down or sit and relax.

5. View the vivid colors [the green grass, the blue sky] that surround you.

6. Take in the sounds around you [birds singing, woodpeckers knocking, crickets chirping]. If you are surrounded by the sounds of war, try to identify a correlating sound that gives you peace, i.e.: a thunder storm, Fourth of July celebration, etc. instead of bombs and guns.

7. Identify the smells of your safe place [fragrant flowers blooming, pine trees, and fresh air].

8. Identify the tastes associated with your safe place [taste the fresh crisp air and honeysuckle].

9. Get in touch with the sensations in your safe place [feel the cool breeze of the shade trees and the warmth of the sun beating through the branches].

10. While in your safe place you can bring in family members, friends, etc. or just be alone and relax. You can also bring in people you are angry with and discuss your issues with them or bring in lost friends and say your goodbyes.

11. Remain as long as you want or are able to.

12. Before you leave your safe place, look around and make sure it looks safe to come back to. Make any adjustments necessary.

13. When you feel comfortable, follow the path back from your safe place and open your eyes slowly.

APPENDIX E
MEDITATION
(Copeland)

1. Select a position that is comfortable for you.

2. Close your eyes or focus on a particular spot on the floor, wall, etc.

3. Spend a few minutes getting in touch with yourself. With your eyes closed, focus on the places where your body touches the chair, cushion, or floor. Notice what this feels like. Now notice those places where one body part touches another. Pay attention to the sensations at these places of contact. Notice how much space your body takes up. Feel the boundary between your body and the space around it.

4. Take several deep breaths and notice your breathing. Notice whether your breathing is fast or slow, deep or shallow, and where your breath goes in your body (high up in your chest, near your stomach, or down low in your abdomen). Now practice moving your breath from one place to another, breathing first into your chest, then down into your stomach area, then down into the lower parts of your torso. Notice your abdomen expanding and contracting. Deep breathes are the most relaxing ones to use when meditating. This may be hard at first, but it will become easier as you practice.

5. Maintain a passive attitude when meditating. Remember that you will have many intrusive thoughts when you first begin to meditate,

but your moments of fixed attention will increase as your ability to let go of stray thoughts improves. Don't worry about whether you are doing things correctly or well enough.   Realize that whatever happened is what is supposed to happen.

6. Let go of intrusive thoughts by taking several deep breaths. As you have a thought or perception, imagine that you are enclosing that thought or perception into a bubble. Then just watch the bubble float away. You may use other images like puffs of smoke or leaves floating down a stream.

APPENDIX F
WORKING WITH NEGATIVE FEELINGS

Stuffing your feelings can become destructive over time. The following is a way of processing your feelings instead of stuffing them. This exercise can be drawn on paper or in your mind with your eyes closed.

1. Draw an outline of your body.

2. Now search your body and identify the location(s) of the feeling (Remember, feelings are one word, i.e., happy, sad, angry, depressed, rejected, abandoned, etc.). Really focus.

3. Use your five senses and:

    a. Draw a picture of what the feeling looks like (image, color, shape, and size)

    b. Give the feeling a taste

    c. Give the feeling a sound

    d. Give the feeling a smell

    e. Describe the texture if you could touch it.

4. Ask:

    a. What is its purpose?

    b. What does it want?

    c. What does it need?

    d. What are its destructive traits?

    e. What are its good traits?

    f. Is there another feeling that helps this feeling do well?

5. Visualize destroying the image of the feeling in a way of your choosing OR give the feeling permission to exist but not to control you.

6. Focus on its good traits and use it for your good.

APPENDIX G
GROUNDING TECHNIQUES

1. Look around and notice details of your surroundings. Touch things.

2. Picture a safe place in your mind using all five senses (Appendix D).

3. Visualize putting overwhelming memories into a strong container or floating them out to sea in a bottle, etc.

4. Distract yourself: take a walk, read a book, listen to the radio/TV, talk with peers, eat something, remember a pleasant smell, hold an ice cube in your hand.

5. Hold something cold against your face like a cold washcloth, a can of chilled soda, ice cube etc.

6. Concentrate intensely: count backwards from 1,000, do math problems in your head or on paper.

7. Yell if necessary. This can be done with your mouth closed so it sounds loud inside your head but like a whimper to others.

8. Journal (See Appendix A).

9. Do something safe but physical to process anger and decrease adrenalin (see chapter on anger).

10. Do breathing and/or relaxation exercises (See Appendix B & C).

11. Use self-hypnosis/Guided Imagery (See Appendix D).

12. Visualize a huge "stop" sign.

13. Use ammonia snaps if you have them.

14. Use positive affirmations.

15. Connect with the here and now.

16. Monitor self talk – change negatives to positives

17. Look in a mirror and smile.

18. Dance.

19. Repeat a Grounding Phrase – "I'm here right now."

20. Pray.

21. Exercise.

22. Work through any anger.

23. Get up and move around if you can. Jump up and down or stretch.

24. Take a shower.

25. Draw.

26. Eat something very slowly and try to identify the ingredients in the food.

27. Find a safe person.

28. **IDENTIFY THE TRIGGER:** The date of some anniversary, the sights, the sounds, the smells, etc. Avoid these triggers as much as possible by changing **who** you are with, **what** you are doing, and/or **where** you are at.

29. Practice these techniques until you find ones that work for you. Highlight those techniques and refer to them in time of need.

30. Create a "safety plan" that works for you.

APPENDIX H
NEED FOR MEDICATION

Hopefully the activities listed in this booklet will help you maintain good mental health. But when nothing else seems to work, you need to consider the need for a medication evaluation. I know that those trying to advance in the military are afraid that taking medications will affect their promotion potential. However, you need to weigh all the factors. If your symptoms are severe, can you function appropriately? If not, your career may be in jeopardy anyway.

Whenever we are faced with severe stressors lasting a great deal of time, or whenever we feel like we have lost control over our lives, we are prone to depression. There are those who state they do not want to take any substance that controls them. Most of the newer medications that are used for mental health purposes do not control you, but help you to be more in control of yourself and your actions. Some medications like antidepressant medications do not make you feel high. They correct a chemical imbalance and help you to feel more like yourself again.

Up until 20 years ago, treatment for mental health was brutal and barbaric. Taking a pill is nothing compared to these old therapies and the sufferings they created. Also, many mental illnesses are genetic and can be triggered by stress. Using this book may help delay that gene from activating, but if it doesn't, you have to consider the need for medication.

Some benefits of taking the correct type and dose of medication are:

1. They can reduce the false or strange beliefs and ideas that are not shared by others.

2. They can decrease tension, agitation – make you calmer and more relaxed.

3. They can help you think clearly and concentrate better. Thoughts that are hostile, strange, or aggressive don't occur as often.

4. They can reduce fears, confusion and insomnia.

5. They can help you feel happier, brighter, and healthier.

6. They can help you act more appropriately; you don't want to laugh, cry, or smile for no reason at all.

Each medication has its own purpose. Don't hesitate to ask your doctor for this information. Anti-psychotic medications are not addictive, even after taking them for many years. Although not addictive, these medications can increase your appetite and can cause weight gain. Regular exercise and eating a proper diet can alleviate this problem. You should not drink alcohol while on these medications.

APPENDIX I
ONLINE RESOURCES

The following is a central site with a wealth of information on mental health. You can use the rest of the page for writing in your own personalized sites.

http://www.psychcentral.com (Grohol)

REFERENCES

Copeland, Mary Ellen (1992). *The Depression Workbook: A Guide for Living with Depression and Manic Depression.* Brattleboro, Vermont: Peach Press.

Florida Hypnosis and Counseling Center, Inc. *Deep Breathing.* Sarasota, FL. Author.

Grohol, John M. (2000). *The Insider's Guide to Mental Health Resources Online.* New York, NY: The Guilford Press.

Horowitz, M.J. (1985). Disasters and Psychological Responses to Stress. *Psychiatric Annual,* 15:161.

Jongsma, Arthur E. Jr., & Peterson, Mark (1995). *The Complete Psychotherapy Treatment Planner.* New York, NY: John Wiley and Sons, Inc.

Kilner, Major Peter G. (2005). *The Military Ethicist's Role in Preventing and Treating Combat-related, Perpetration-Induced Psychological Trauma.* DRAFT.

Kolski, Tammi D., Avriette, Michael, & Jongsma Jr., Arthur (2001). *The Crisis Counseling and Traumatic Events Treatment Planner.* New York, NY: John Wiley and Sons, Inc.

Peurifoy, Reneau Z. (2005). *Anxiety, Phobias and Panic,* (Rev. ed.). New York, NY: Warner Books.

ABOUT THE AUTHOR

Ginny Sartini is an experienced counselor who has worked in various treatment settings since getting her Master's degree in 1990. She has worked for several psychiatric hospitals and on a women's trauma unit in Central Florida. Ms. Sartini was in private practice in Florida for 10 years and now works for the Tulsa Outpatient Treatment Center of Veterans Affairs in Tulsa, Oklahoma.

She pursued her Bachelor's Degree in Business Administration at Upper Iowa University, in Fayette, Iowa and a Master's Degree in Social Work at the University of Kentucky, in Lexington, Kentucky. She has taken several courses in Post Traumatic Stress Disorder and other psychiatric diagnoses and treatment interventions.

Ginny Sartini is licensed by the Florida Board of Clinical Social Workers and is listed in the NASW register.